Editor
Lorin Klistoff, M.A.

Editorial Manager
Karen J. Goldfluss, M.S. Ed.

Editor in Chief
Sharon Coan, M.S. Ed.

Cover Artist
Jessica Orlando

Art Coordinator
Denice Adorno

Creative Director
Elayne Roberts

Imaging
Stephanie A. Salcido
James Edward Grace

Graphic Illustrator
James Edward Grace

Product Manager
Phil Garcia

Publisher
Mary D. Smith, M.S. Ed.

How to Work with Fractions

Grades 2–3

Author

Mary Rosenberg

Teacher Created Resources

Teacher Created Resources, Inc.
6421 Industry Way
Westminster, CA 92683
www.teachercreated.com
ISBN-1-57690-954-9
©2000 Teacher Created Resources, Inc.
Reprinted, 2005
Made in U.S.A.

Table of Contents

How to ···················· Use This Book

A Note to Teachers and Parents

Welcome to the "How to" math series! You have chosen one of over two dozen books designed to give your children the information and practice they need to acquire important concepts in specific areas of math. The goal of the "How to" math books is to give children an extra boost as they work toward mastery of the math skills established by the National Council of Teachers of Mathematics (NCTM) and outlined in grade-level scope and sequence guidelines. The NCTM standards encourage children to learn basic math concepts and skills and apply them to new situations and real-world events. The children learn to justify their solutions through the use of pictures, words, graphs, charts, and diagrams.

The design of this book is intended to allow it to be used by teachers or parents for a variety of purposes and needs. Each of the units contains one or more "How to" pages and two or more practice pages. The "How to" section of each unit precedes the practice pages and provides needed information such as a concept or math rule review, important terms and formulas to remember, or step-by-step guidelines necessary for using the practice pages. While most "How to" pages are written for direct use by the children, in some lower-grade level books these pages are presented as instructional pages. In this book, the "How to" page details the concepts that will be covered in the pages that follow as well as how to teach the concepts. Many of the "How to" pages also include "Learning Notes." The practice pages review and introduce new skills and provide opportunities for the children to apply the newly acquired skills. Each unit is sequential and builds upon the ideas covered in the previous unit(s).

About This Book

How to Work with Fractions: Grades 2–3 presents a comprehensive overview of fractions for students at this level. It can be used to introduce and teach basic fractions to children with little or no background in the concepts. The units in this book can be used in whole-class instruction with the teacher or by a parent assisting his or her child through the book. This book also lends itself to use by a small group in earlier grades engaged in enrichment or accelerated work. A teacher may want to have two tracks within his or her class with one moving at a faster pace and the other at a gradual pace appropriate to the ability or background of the children. This book can also be used in a learning center containing materials needed for each unit of instruction.

When working with fractions, it is important to give children many concrete experiences using manipulatives as often as possible. The more experiences the children have with manipulatives, the easier it will be for them to develop their understanding of fractions.

The activities in this book will help your children learn new skills or reinforce skills already learned in several areas. The activities explain the vocabulary of fractions, demonstrate mixed and whole numbers, and include the addition and multiplication of fractions. This book also integrates fractions into word problems and the computer. It introduces congruent parts, equivalent fractions, and the comparison of fractions.

If children have difficulty with a specific concept or unit within this book, review the materials and allow them to redo the troublesome pages. Since concept development in these units is sequential, it is not advisable to skip units. It is preferable for children to find the work easy at first and to gradually advance to more difficult concepts.

The units in this book are designed to match the suggestions of the National Council of Teachers of Mathematics (NCTM). They strongly support the learning of fractions and other processes in the context of problem solving and real-world applications. Use every opportunity to have students apply these new skills in classroom situations and at home. This will reinforce the value of the skill as well as the process. This book matches a number of NCTM standards including these main topics and specific features:

Problem Solving

The children develop and apply strategies to solve problems, verify and interpret results, sort and classify objects, and solve word problems.

Communication

The children are able to communicate mathematical solutions through manipulatives, pictures, diagrams, numbers, and words. Children are able to relate everyday language to the language and symbols of math. Children have opportunities to read, write, discuss, and listen to math ideas.

Reasoning

Children make logical conclusions through interpreting graphs, patterns, and facts. The children are able to explain and justify their math solutions.

Connections

Children are able to apply math concepts and skills to other curricular areas and to the real world.

Number Sense and Numeration

Children learn to count, label, and sort collections as well as learn the basic math operations of fraction addition and subtraction.

Concepts of Whole Number Operations

Children develop an understanding of the operations (addition and subtraction) by modeling and discussing situations relating math language and the symbols of operation (+ and −) to the fraction problem being discussed.

Other Standards

Children work toward **whole number computation** mastery as they model, explain, and develop competency in basic facts, mental computation, and **estimation** techniques.

Children explore **geometry** and develop **spatial sense** by describing models, drawing and classifying shapes, and relating geometric ideas to number and measurement ideas.

Children learn about **statistics** and **probability** by collecting and organizing data into graphs, charts, and tables.

Children develop concepts of **fractions** through the use of pattern blocks.

Learning Notes

In this unit, children will learn to write the fraction (numerator and denominator) for specific shapes.

Materials

- ten 3" x 5" (8 cm x 13 cm) index cards for each child
- crayons or markers

Teaching the Lesson

Introduce the children to the terms *numerator* and *denominator*. Explain to them that the top number is called the *numerator*. The numerator tells how many parts of the whole are needed or are used. The bottom number is called the *denominator*. The denominator tells how many equal parts there are in all.

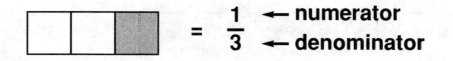

Give each child 10 index cards with a circle or a square on each card and each shape divided into 1 to 10 congruent parts. (Congruent parts are the same size and shape.) Using markers or crayons, have the children color in one part of each shape. Model for the children how to write the fraction for each shape on the index cards. (See the samples below.)

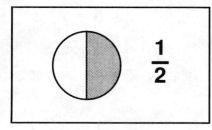 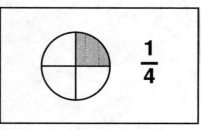 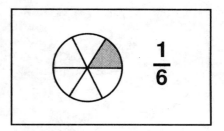

Name a specific fraction and have the children show you their matching index card. For example, "Show me $\frac{1}{4}$." The children would hold up a card that shows $\frac{1}{4}$.

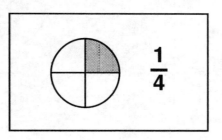

A *fraction* is a part (or parts) of a whole item or shape. When writing fractions, the top number is called the *numerator*. The numerator tells how many parts of the whole are needed or are used. The bottom number is called the *denominator*. The denominator tells how many equal parts there are in all.

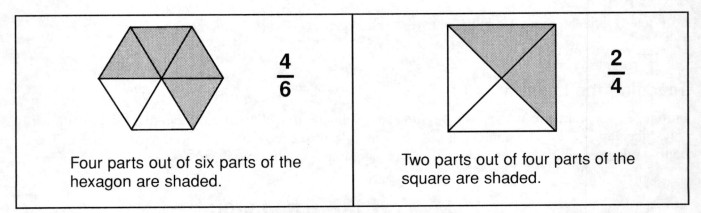

Four parts out of six parts of the hexagon are shaded. $\frac{4}{6}$

Two parts out of four parts of the square are shaded. $\frac{2}{4}$

Directions: Look at the shapes below. Parts of each shape are shaded. Write the numerator for each one of the fractions. The first one has already been done for you.

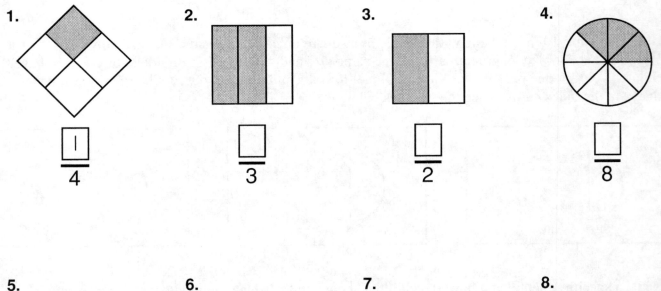

1. $\frac{1}{4}$

2. $\frac{}{3}$

3. $\frac{}{2}$

4. $\frac{}{8}$

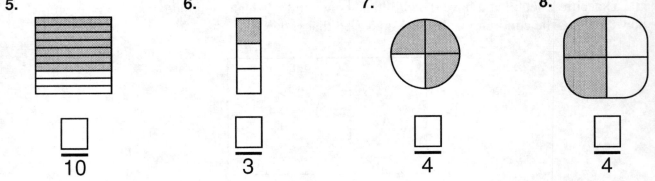

5. $\frac{}{10}$

6. $\frac{}{3}$

7. $\frac{}{4}$

8. $\frac{}{4}$

A *fraction* is a part (or parts) of a whole item or shape. When writing fractions, the top number is called the *numerator*. The numerator tells how many parts of the whole are needed or are used. The bottom number is called the *denominator*. The denominator tells how many equal parts there are in all.

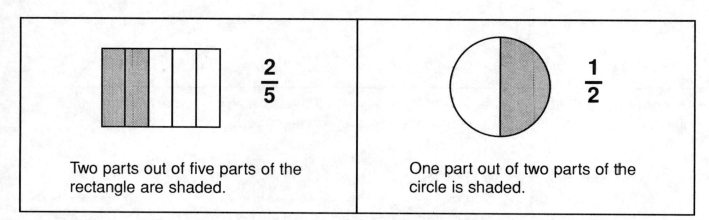

Two parts out of five parts of the rectangle are shaded.

One part out of two parts of the circle is shaded.

Directions: Look at the shapes below. Parts of each shape are shaded. Write the denominator for each one of the fractions. The first one has already been done for you.

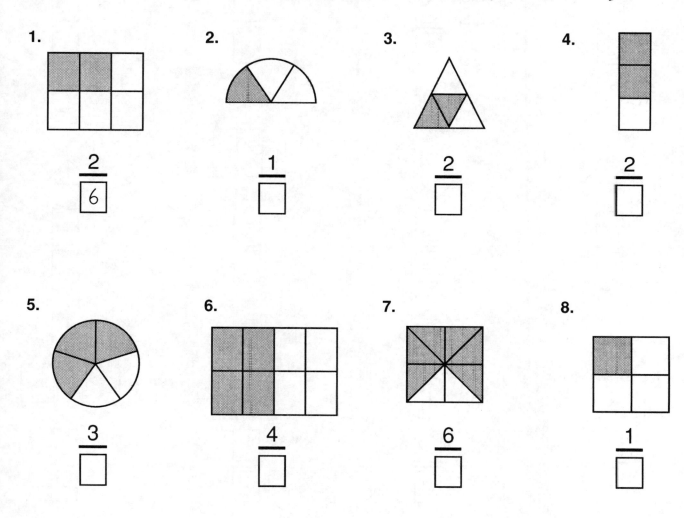

1.

$\dfrac{2}{6}$

2.

$\dfrac{1}{\Box}$

3.

$\dfrac{2}{\Box}$

4.

$\dfrac{2}{\Box}$

5.

$\dfrac{3}{\Box}$

6.

$\dfrac{4}{\Box}$

7.

$\dfrac{6}{\Box}$

8.

$\dfrac{1}{\Box}$

A *fraction* is a part (or parts) of a whole item or shape.

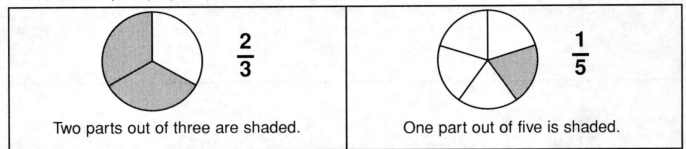

$\dfrac{2}{3}$	$\dfrac{1}{5}$
Two parts out of three are shaded.	One part out of five is shaded.

Directions: Look at each shape. Write the fraction that tells how many parts of the whole object are shaded. The first one has already been done for you.

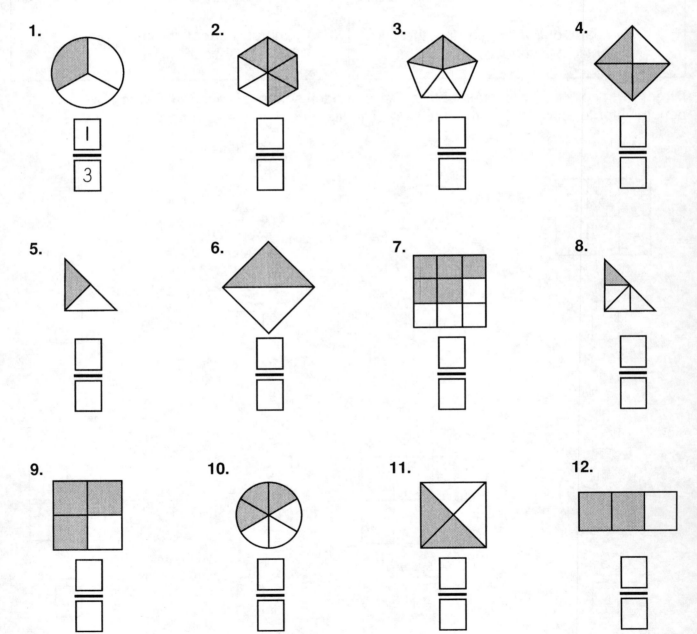

1. $\dfrac{1}{3}$

2.

3.

4.

5.

6.

7.

8.

9.

10.

11.

12.

Learning Notes

In this unit, children recognize shapes with congruent parts. They will also count the number of congruent parts and record the answer. The children will divide shapes into a specific number of congruent parts and will record the fraction.

Materials

- 1 paper plate for each child
- scissors
- pencil or marker for labeling

Teaching the Lesson

1. Have the children fold the paper plate in half.

2. Open the paper plate and cut on the folded line.

3. Label one of the pieces $\frac{1}{2}$ (as shown below).

4. Take the other piece and fold it in half.

5. Open the piece and cut on the folded line.

6. Label one of the pieces $\frac{1}{4}$ (as shown below).

7. Continue the above steps using the unlabeled piece of the plate to make fraction pieces showing $\frac{1}{8}$, $\frac{1}{16}$, and 2 pieces labeled $\frac{1}{32}$.

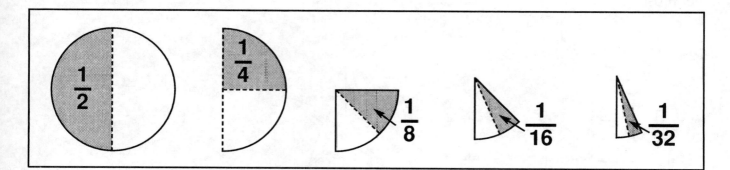

Activity

- Have the children show $\frac{1}{4}$ of the paper plate. Ask, "How many $\frac{1}{4}$ pieces are needed to make a whole paper plate?" *(4)*

- Have the children show $\frac{1}{16}$ of the paper plate. Ask, "How many $\frac{1}{16}$ pieces are needed to make a whole paper plate?" *(16)*

- Continue in this manner until the children feel comfortable working with the different fractions.

Look at the shapes below. All of the shapes are divided into 2 equal parts. When all of the parts are the same size and shape, they are called *congruent parts*.

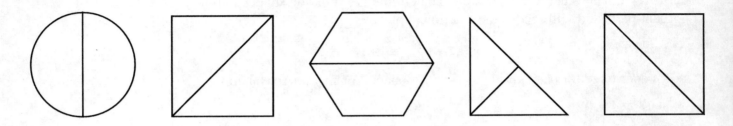

Directions: Look at all of the shapes below. All of the shapes have congruent parts. Count the number of congruent parts in each shape. Write the number on the line.

1.

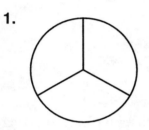

There are _____ congruent parts.

2.

There are _____ congruent parts.

3.

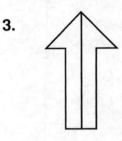

There are _____ congruent parts.

4.

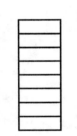

There are _____ congruent parts.

5.

There are _____ congruent parts.

6.

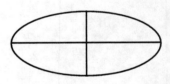

There are _____ congruent parts.

7.

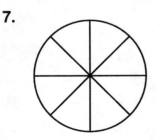

There are _____ congruent parts.

8.

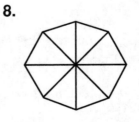

There are _____ congruent parts.

9.

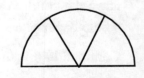

There are _____ congruent parts.

2 ⟩ **Practice** · · · · · · · · · · · · **Same Size, Same Shape**

When a shape is divided equally, each piece of the shape is the same size and same shape. Look at the two squares below. One square has been divided into four congruent parts. Each part is the same size and shape. One square has been divided into four parts, but the parts are not congruent. The parts are not the same size or shape.

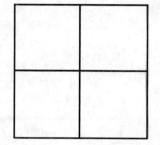

This square has congruent parts.

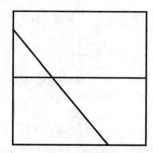

This square does not have congruent parts.

Directions: Look at each pair of shapes. Circle the one shape that has congruent parts.

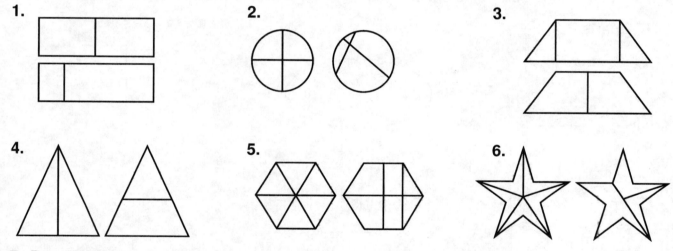

1.

2.

3.

4.

5.

6.

7. Draw 2 shapes. Divide one of the shapes into parts that are the same size and shape. Divide the other shape into parts that are not the same size and shape.

8. What does "congruent" mean? _____

_____

_____

When a shape is divided into congruent parts, each part is a fraction of the shape. Follow the steps below.

1. Divide the shape into congruent parts.

2. Shade one part of the shape.

3. Next, write the fraction. When writing the fraction, the top number is called the *numerator*. The numerator tells the number of each piece (or pieces) used or needed. The bottom number is called the *denominator*. The denominator tells how many equal pieces there are in all. The fraction for one part of this rectangle is $\frac{1}{3}$.

3 equal parts

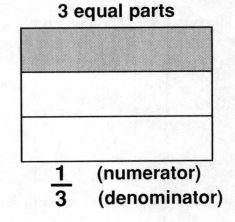

$\frac{1}{3}$ **(numerator)**
       (denominator)

Directions: Divide each shape into congruent parts. Shade one part of the shape. Write the fraction for one part of the shape. The first one has already been done for you.

1. 2 parts

2. 4 parts

3. 3 parts

4. 4 parts

5. 6 parts

6. 5 parts

7. 8 parts

8. 4 parts

9. 3 parts

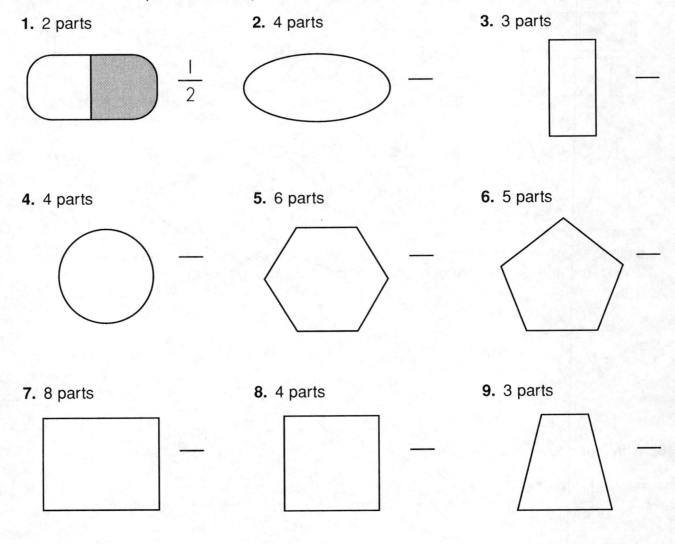

Learning Notes

In this unit, children will write the fractions that describe different sets of items and use addition to check their answers.

Materials

- teddy bear counters (or beans, pennies, multilink cubes, blocks, etc.)
- chalkboard, dry erase board, or piece of paper to record different fractions

Teaching the Lesson

Place 12 teddy bear counters on a table. Write "$\frac{1}{2}$ of 12" on the chalkboard. Ask the children, "What does $\frac{1}{2}$ of 12 mean?" (*2 groups of equal size*) Have the children divide the teddy bear counters into 2 groups of equal size (2 groups of 6 teddy bear counters). Write "$\frac{1}{2}$ of 12 = 6 teddy bear counters." Use addition to check the answer. Add the first group of teddy bear counters with the second group of teddy bear counters. The total number of teddy bear counters is 12.

Using the teddy bear counters, have the children find the answers for the following fractions:

$\frac{1}{3}$ of 12 *(4)*	$\frac{1}{3}$ of 15 *(5)*	$\frac{1}{2}$ of 20 *(10)*
$\frac{1}{4}$ of 12 *(3)*	$\frac{1}{5}$ of 15 *(3)*	$\frac{1}{4}$ of 20 *(5)*
$\frac{1}{6}$ of 12 *(2)*	$\frac{1}{2}$ of 14 *(7)*	$\frac{1}{5}$ of 20 *(4)*

Fractions can be used to describe sets of items. Look at the example below. There are 6 pigs. One half of the pigs in the line is circled. Now, there are 2 groups of pigs.

There are 6 pigs. One half of 6 is 3. The fraction is $\frac{3}{6}$.

Directions: Circle half of the items in each set below, and then complete the math sentences. Write the fraction for each set of items on the last line. The first one has already been done for you.

1. There are ___8___ bats .

2. One half of ___8___ is ___4___.

3. The fraction is ___$\frac{4}{8}$___.

4. There are _____ _____.

5. One half of _____ is _____.

6. The fraction is _____.

7. There are _____ _____.

8. One half of _____ is _____.

9. The fraction is _____.

10. There are _____ _____.

11. One half of _____ is _____.

12. The fraction is _____.

13. There are _____ _____.

14. One half of _____ is _____.

15. The fraction is _____.

16. There are _____ _____.

17. One half of _____ is _____.

18. The fraction is _____.

Sets of items can be divided into thirds. Look at the example below. There are 9 birds in a row. One third of the birds is circled. You can divide the birds equally into three groups with three birds in each group.

There are 9 birds. One third of 9 is 3. The fraction is $\frac{3}{9}$.

Directions: Circle one third of the items in each set below, and then complete the math sentences. Write the fraction for each set of items on the last line. The first one has already been done for you.

1. There are ____9____ ____bones____. One third of ___9___ is ___3___.

The fraction is ___$\frac{3}{9}$___.

2. There are ____ ____. One third of ____ is ____.

The fraction is ____.

3. There are ____ ____. One third of ____ is ____.

The fraction is ____.

4. There are ____ ____. One third of ____ is ____.

The fraction is ____.

Groups of items can be divided into many different fractions. You can find the fraction for each group by circling the correct number of items. You can check your answer by adding. Look at the example below.

Find $\frac{1}{3}$ of 9 flowers.

$\frac{1}{3}$ of 9 is 3.

To check, add 3 three times. The answer should be 9, the total number of flowers. 3 + 3 + 3 = 9

Directions: Find the fraction set for each group below by circling the correct number of items. Check your answer by adding. The first one has already been done for you. Remember to show your work.

1. $\frac{1}{5}$ of 10 cupcakes = _____2_____

$\frac{1}{5}$ of 10 is 2.

Check answer:
2 + 2 + 2 + 2 + 2 = 10

2. $\frac{1}{2}$ of 10 balloons = _____

3. $\frac{1}{4}$ of 8 ice cream bars = _____

4. $\frac{1}{8}$ of 8 feathers = _____

Learning Notes

In this unit, children will complete a fraction table and will divide shapes into different fractions. They will compare fractions with common denominators and put fraction circles in order from smallest to greatest. The children will also write the fractions for specific shapes.

Materials

- scissors
- crayons or markers
- o-rings or paper fasteners (brads)
- chalkboard, dry erase board, or a piece of paper

Teaching the Lesson

What is the Fraction? (page 18)

Review the meanings of the numerator and the denominator by drawing a square on the chalkboard. Write the fraction "$\frac{1}{4}$." Ask the children, "What does the fraction $\frac{1}{4}$ tell you about the square?" (*The square is divided into 4 parts and 1 of the parts is shaded.*) Repeat the above steps with several other shapes and fractions.

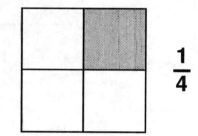

Comparing Fractions (page 19)

On the chalkboard, draw two circles. Color $\frac{2}{3}$ of one circle and $\frac{1}{3}$ of the other circle. Ask the children, "Which circle has the largest amount colored?" ($\frac{2}{3}$) Write the math sentence that describes the two circles ($\frac{2}{3} > \frac{1}{3}$). Repeat the above steps with several other shapes and fractions.

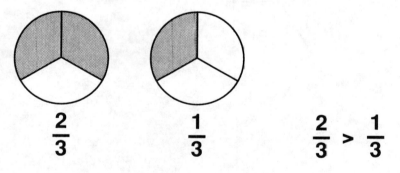

$$\frac{2}{3} \qquad \frac{1}{3} \qquad \frac{2}{3} > \frac{1}{3}$$

Ordering Fractions (page 20)

Have the children color each circle according to its fraction. Cut the fraction cards apart and place them in order, from smallest to greatest, by the amount of space colored. Place the cards on an o-ring or on a paper fastener. The children can use the cards as a reference tool.

Directions: Complete the fraction table below.

	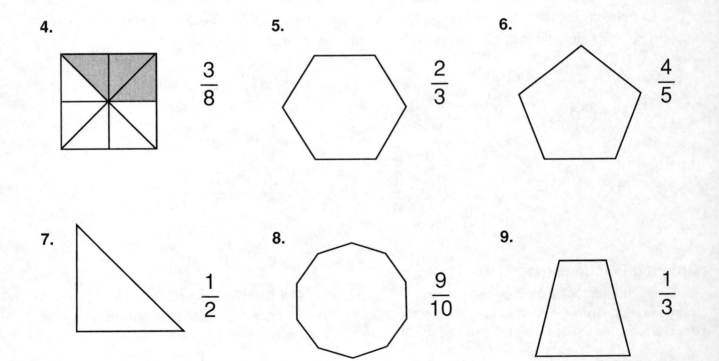					
1. How many parts does each circle have?			**3**		**5**	**6**
2. How much is each part of the shape?	one whole			one fourth		
3. What is the fraction for each part?		$\frac{1}{2}$				

Directions: Divide each shape into the correct number of parts. Shade each shape to show the fraction. The first one has already been done for you.

4. $\frac{3}{8}$

5. $\frac{2}{3}$

6. $\frac{4}{5}$

7. $\frac{1}{2}$

8. $\frac{9}{10}$

9. $\frac{1}{3}$

When comparing fractions with the same denominator, look at the numerator. The fraction with the larger numerator is the larger fraction.

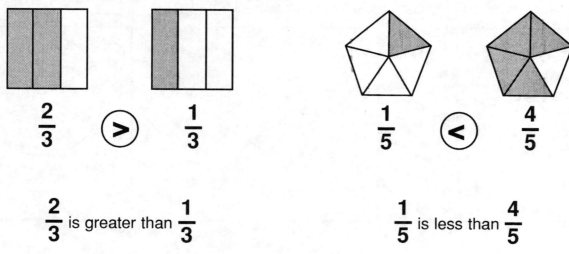

$\frac{2}{3}$ is greater than $\frac{1}{3}$ $\frac{1}{5}$ is less than $\frac{4}{5}$

Directions: Compare the two fractions using the ">" (greater than) and "<" (less than) symbols. Then write the math sentence. The first one has already been done for you.

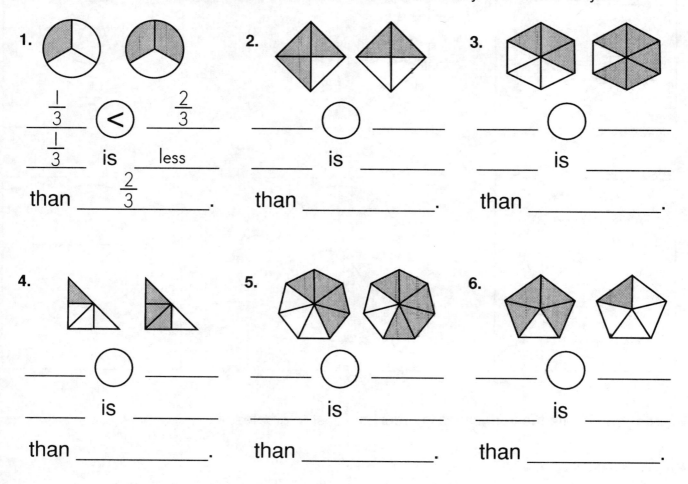

Directions: Color in each circle to show the correct fraction. Cut the fraction cards apart and place them in order from the smallest fraction to the largest fraction. Punch a hole in the corner of each card and put the cards on an o-ring.

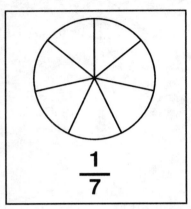

$\dfrac{1}{7}$

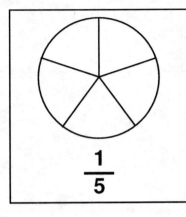

$\dfrac{1}{5}$

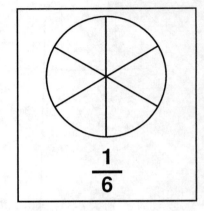

$\dfrac{1}{6}$

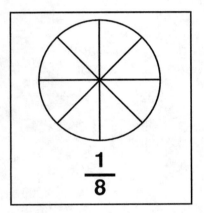

$\dfrac{1}{8}$

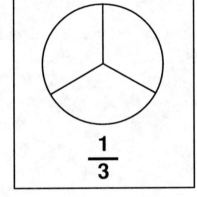

$\dfrac{1}{3}$

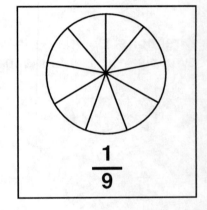

$\dfrac{1}{9}$

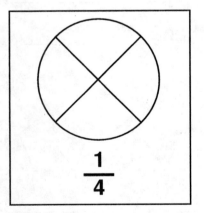

$\dfrac{1}{4}$

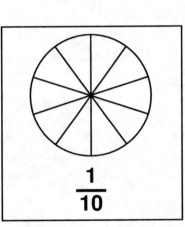

$\dfrac{1}{10}$

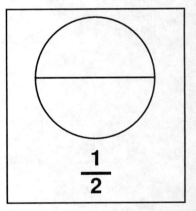

$\dfrac{1}{2}$

Directions: Write these fractions in order, smallest to greatest.

_____, _____, _____, _____, _____, _____, _____, _____, _____

Learning Notes

In this unit, children will read word problems and write the correct mixed number. They will look at different sets of items and write the whole number or mixed number. They will also rewrite mixed numbers.

Materials

- 10 paper plates with one of the plates cut into different fraction amounts ($\frac{1}{2}$, $\frac{1}{4}$, and $\frac{1}{8}$)
- large piece of butcher paper or poster board to make the chart

Teaching the Lesson

Explain the following chart to the children:

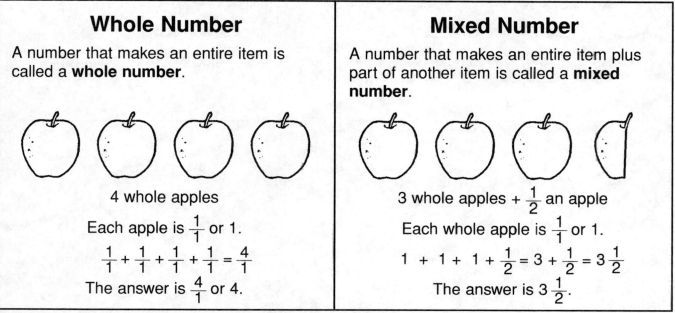

Whole Number	**Mixed Number**
A number that makes an entire item is called a **whole number**.	A number that makes an entire item plus part of another item is called a **mixed number**.
4 whole apples	3 whole apples $+ \frac{1}{2}$ an apple
Each apple is $\frac{1}{1}$ or 1.	Each whole apple is $\frac{1}{1}$ or 1.
$\frac{1}{1} + \frac{1}{1} + \frac{1}{1} + \frac{1}{1} = \frac{4}{1}$	$1 + 1 + 1 + \frac{1}{2} = 3 + \frac{1}{2} = 3\frac{1}{2}$
The answer is $\frac{4}{1}$ or 4.	The answer is $3\frac{1}{2}$.

(**Note:** Fractions must have the same denominator to be added together. When adding fractions with the same denominator, add the numerators together and record the answer. The denominator stays the same. For example, $\frac{1}{2} + \frac{1}{2} = \frac{2}{2}$.)

Using the paper plates, model different whole numbers and mixed numbers. For example, show three paper plates. Each paper plate is $\frac{1}{1}$ or 1. The fraction for three plates is $\frac{3}{1}$ or 3. Show 4 and $\frac{1}{2}$ paper plates. The answer is $4\frac{1}{2}$. Do this several times until the children are comfortable with the concept of whole numbers and mixed numbers.

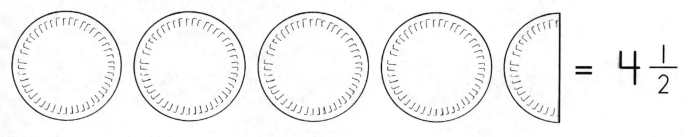

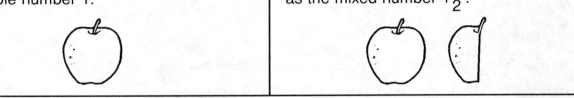

A number that makes an entire item is called a *whole number*. The picture below can be represented as the fraction $\frac{1}{1}$ or the whole number 1.

A number that makes an entire item plus part of another item is called a *mixed number*. The picture below is represented as the mixed number $1\frac{1}{2}$.

Directions: Look at each set of pictures. Circle whether a whole number or a mixed number is shown. Then write the number on the line. The first one has already been done for you.

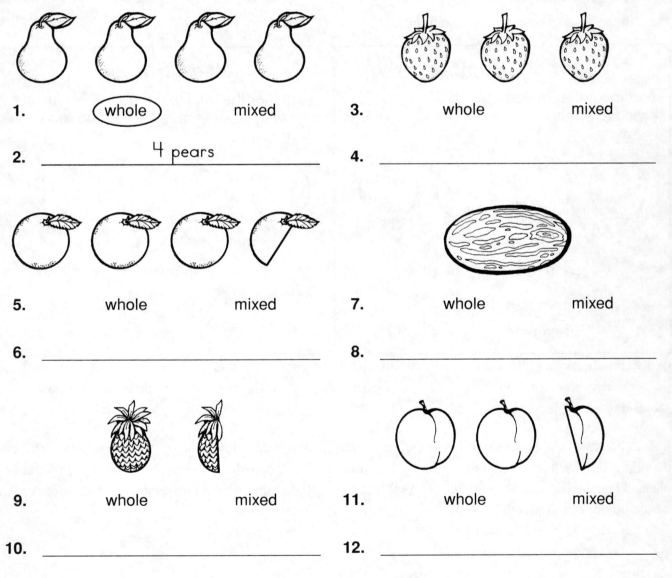

1. (whole)   mixed

2. _____ 4 pears _____

3. whole   mixed

4. _____

5. whole   mixed

6. _____

7. whole   mixed

8. _____

9. whole   mixed

10. _____

11. whole   mixed

12. _____

13. Write these whole numbers and mixed numbers in order from smallest to greatest.

_____, _____, _____, _____, _____, _____

A *mixed number* means that a whole item plus a part of another item is needed or was used.

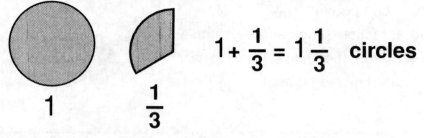

$$1 + \frac{1}{3} = 1\frac{1}{3} \quad \text{circles}$$

Directions: Read each word problem. Draw a picture and write the mixed number to solve each word problem. The first one has already been done for you.

1. Brianna was really hungry when she got home from school. She ate a whole sandwich and half of another sandwich. How many sandwiches did Brianna eat?

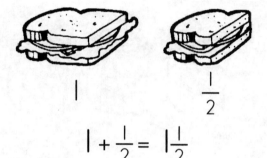

$$1 + \frac{1}{2} = 1\frac{1}{2}$$

Brianna ate _____ $1\frac{1}{2}$ _____ sandwiches.

2. Lorenzo found three pairs of socks and one single sock. How many pairs of socks did Lorenzo find?

Lorenzo found _____ pairs of socks.

3. Dominique has one whole candy bar and one third of another candy bar. How many candy bars does Dominique have?

Dominique has _____ candy bars.

4. Gabriel ate one whole pepperoni pizza and one fourth of another pizza. How many pizzas did Gabriel eat?

Gabriel ate _____ pizzas.

Mixed numbers can be written with whole numbers and a fraction. Look at the example. There are 3 whole muffins and one half of a muffin. The written mixed number is $3\frac{1}{2}$.

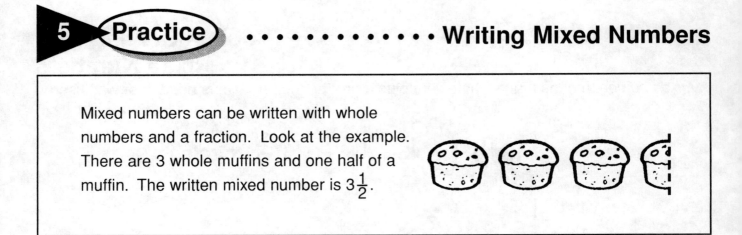

Directions: Write a mixed number to represent each set of pictures. The first one is done for you.

1.
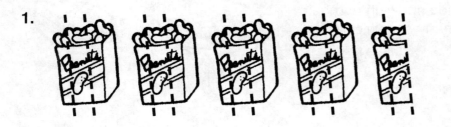
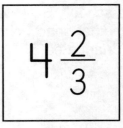

$4\frac{2}{3}$

2.
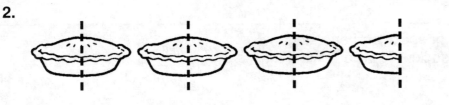

3.
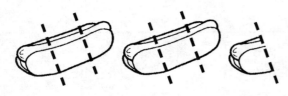

4.

5. Write the above mixed numbers in order from smallest to greatest.

_____, _____, _____, _____,

Learning Notes

In this unit, children will learn to find equivalent fractions by multiplying the numerator and the denominator by the same number. They will find equivalent fractions by shading two shapes to cover the same amount of area. The equivalent fractions will be recorded. The children will also check two fractions for equivalency through the use of multiplication.

Materials

- a piece of paper with 10 circles drawn on it (Each circle will show a different fraction. See samples below. Enlarge pictures as you see fit.)

- scissors

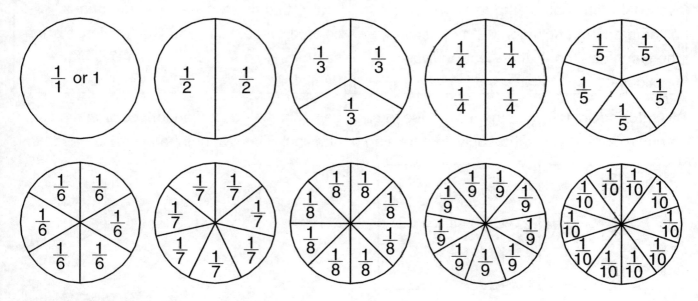

Teaching the Lesson

Have children cut the circles out and then cut each circle into its different pieces. Have the children place the $\frac{1}{2}$ circle on their tables. Then, show $\frac{1}{2}$ using the other fraction pieces. Have the children also find different ways of showing $\frac{1}{3}$, $\frac{1}{4}$, $\frac{2}{3}$, etc., using the fraction circles.

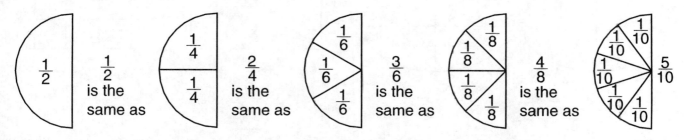

Review with the children how to multiply to find equivalent fractions. Remind the children to multiply the numerators together and to multiply the denominators together. The fraction must be multiplied by the same number to get an equivalent fraction.

$$\frac{2}{4} \times \frac{2}{2} = \frac{(2 \times 2)}{(4 \times 2)} = \frac{4}{8}$$

Fractions that name the same amount of a set of items are called *equivalent fractions*.

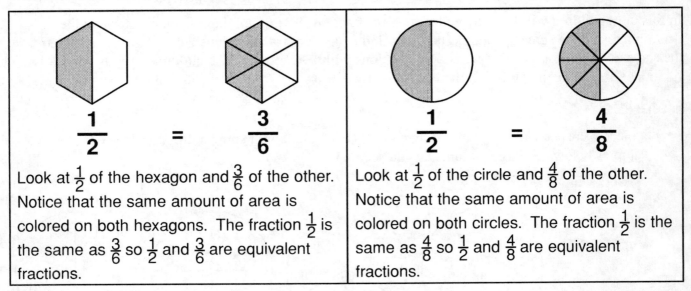

$$\frac{1}{2} = \frac{3}{6}$$

Look at $\frac{1}{2}$ of the hexagon and $\frac{3}{6}$ of the other. Notice that the same amount of area is colored on both hexagons. The fraction $\frac{1}{2}$ is the same as $\frac{3}{6}$ so $\frac{1}{2}$ and $\frac{3}{6}$ are equivalent fractions.

$$\frac{1}{2} = \frac{4}{8}$$

Look at $\frac{1}{2}$ of the circle and $\frac{4}{8}$ of the other. Notice that the same amount of area is colored on both circles. The fraction $\frac{1}{2}$ is the same as $\frac{4}{8}$ so $\frac{1}{2}$ and $\frac{4}{8}$ are equivalent fractions.

Directions: Color in $\frac{1}{2}$ of the first shape in each pair. On the second shape, color in the same amount of area. Write the equivalent fraction on the line. The first one has been done for you.

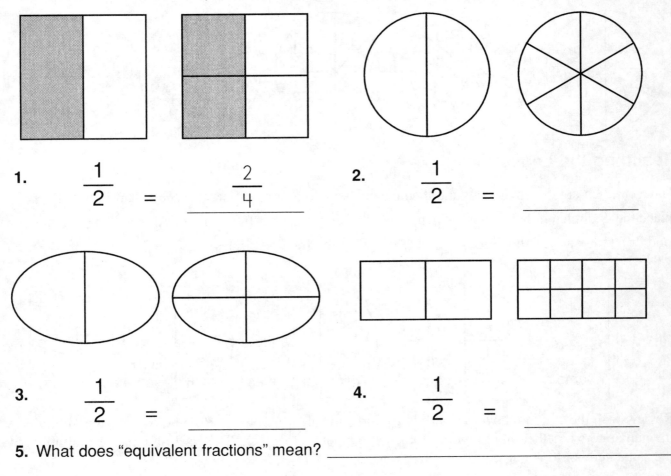

1. $\frac{1}{2}$ = $\frac{2}{4}$ 2. $\frac{1}{2}$ = _____

3. $\frac{1}{2}$ = _____   4. $\frac{1}{2}$ = _____

5. What does "equivalent fractions" mean? _____

_____

Multiplication can be used to find equivalent fractions. Multiply the numerator (top number) and the denominator (bottom number) by the same number.

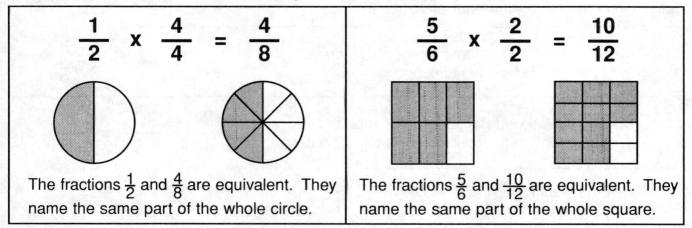

$$\frac{1}{2} \times \frac{4}{4} = \frac{4}{8}$$

The fractions $\frac{1}{2}$ and $\frac{4}{8}$ are equivalent. They name the same part of the whole circle.

$$\frac{5}{6} \times \frac{2}{2} = \frac{10}{12}$$

The fractions $\frac{5}{6}$ and $\frac{10}{12}$ are equivalent. They name the same part of the whole square.

Directions: Multiply to find equivalent fractions for each one of the fractions. Write the equivalent fraction in the space provided. The first one has already been done for you.

1.
$$\frac{8}{9} \times \frac{2}{2} = \frac{16}{18}$$

2.
$$\frac{3}{7} \times \frac{8}{8} = \underline{}$$

3.
$$\frac{7}{8} \times \frac{3}{3} = \underline{}$$

4.
$$\frac{8}{10} \times \frac{4}{4} = \underline{}$$

5.
$$\frac{1}{2} \times \frac{7}{7} = \underline{}$$

6.
$$\frac{3}{6} \times \frac{5}{5} = \underline{}$$

7.
$$\frac{6}{7} \times \frac{6}{6} = \underline{}$$

8.
$$\frac{4}{5} \times \frac{9}{9} = \underline{}$$

9.
$$\frac{2}{10} \times \frac{4}{4} = \underline{}$$

10. What happens when a fraction is multiplied by 1?_____

Multiplication is used to see if two fractions are equivalent (or equal). Multiply the fractions so that they both have the same denominator. Then check for equivalency. Look at the example below to see if $\frac{1}{3}$ and $\frac{2}{6}$ are equivalent.

Multiply both the numerator and the denominator by the other fraction's denominator.	$\frac{1}{3} \times \frac{6}{6} = \frac{6}{18}$
Multiply the second fraction's numerator and denominator by the other fraction's denominator.	$\frac{2}{6} \times \frac{3}{3} = \frac{6}{18}$
The products are the same. They both equal $\frac{6}{18}$ so $\frac{1}{3}$ and $\frac{2}{6}$ are equivalent fractions.	$\frac{1}{3} = \frac{2}{6}$

Directions: Color in each shape to show the fraction. Use multiplication to check for equivalency. If the fractions are equivalent, circle "yes." If the fractions are not equivalent, circle "no." Remember to show your work. The first one has already been done for you.

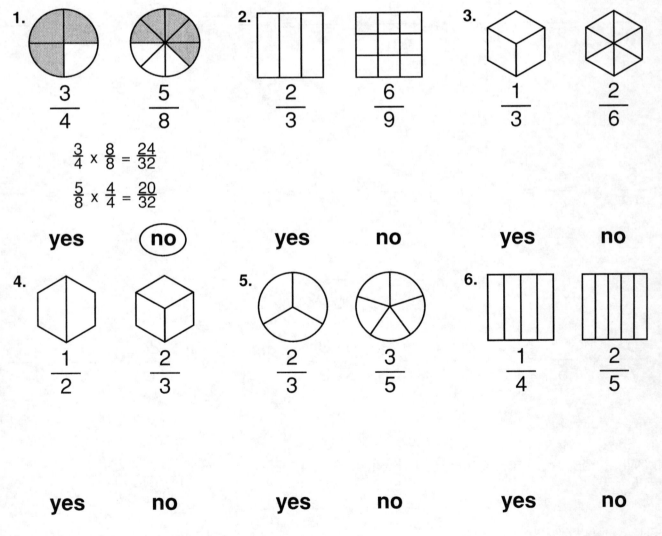

1. $\frac{3}{4}$ $\frac{5}{8}$

$\frac{3}{4} \times \frac{8}{8} = \frac{24}{32}$

$\frac{5}{8} \times \frac{4}{4} = \frac{20}{32}$

yes   (**no**)

2. $\frac{2}{3}$ $\frac{6}{9}$

yes   **no**

3. $\frac{1}{3}$ $\frac{2}{6}$

yes   **no**

4. $\frac{1}{2}$ $\frac{2}{3}$

yes   **no**

5. $\frac{2}{3}$ $\frac{3}{5}$

yes   **no**

6. $\frac{1}{4}$ $\frac{2}{5}$

yes   **no**

Learning Notes

In this unit, children will divide to find the number of items for a specific fraction and then use multiplication to check the answers. They will also solve various fraction word problems.

Materials

- scratch paper to use when solving the math problems
- pencils

Teaching the Lesson

Show the children the steps for using division to find the total number of items for a specific fraction.

Step 1: Read the problem.	**Step 2:** Divide the dividend (15) by the divisor (the denominator). Read the answer.	**Step 3:** Check the answer by multiplying the quotient (5) by the divisor (3).
1/3 of 15	**15 ÷ 3 = 5** **1/3 of 15 = 5**	**5 x 3 = 15**

For more complex problems, show the children how to use both division and multiplication to solve the problem.

Step 1: Read the problem.	**Step 2:** Divide the dividend (18) by the divisor (the denominator).	**Step 3:** Multiply the quotient (6) by the numerator (2).
2/3 of 18	**18 ÷ 3 = 6**	**6 x 2 = 12** **2/3 of 18 is 12**

For the word problems, model how to read the problem and use the information to make the math problem.

Step 1: Read the word problem and circle the important information. Jason had (24 marbles.) He gave (2/3) of the marbles to his friend. How many marbles did Jason give away?	**Step 2:** Solve the problem. **24 ÷ 3 = 8** **8 x 2 = 16** Jason gave away 16 marbles.

7 ▶ Practice · · · · · Using Division to Find the Fraction

Division can be used to find the fraction for larger groups of items. Divide the denominator (divisor) into the total number of items in the set (dividend). The answer will be the number of items or the size of each part (quotient) in the fraction.

What is $\frac{1}{3}$ of 12 apples?

12 ÷ 3 = 4
(dividend) ÷ (divisor) = (quotient)

There are 3 groups of apples (the denominator).
There are 4 apples in each group ($\frac{1}{3}$ of 12).

To check your division, multiply the answer (quotient) by the denominator (divisor). The answer should equal the total number of items in each set.

4 x 3 = 12

(quotient) x (divisor) = (dividend)

Directions: Use division to find the fraction for each number and circle the amount in the picture. Use multiplication to check your answer. The first one has already been done for you.

---

1. $\frac{1}{4}$ of 16 buttons

$\underline{\quad 16 \quad}$ ÷ $\underline{\quad 4 \quad}$ = $\underline{\quad 4 \quad}$ buttons

$\underline{\quad 4 \quad}$ x $\underline{\quad 4 \quad}$ = $\underline{\quad 16 \quad}$

2. $\frac{1}{3}$ of 15 stars

$\underline{\qquad}$ ÷ $\underline{\qquad}$ = $\underline{\qquad}$ stars

$\underline{\qquad}$ x $\underline{\qquad}$ = $\underline{\qquad}$

---

3. $\frac{1}{5}$ of 10 cars

$\underline{\qquad}$ ÷ $\underline{\qquad}$ = $\underline{\qquad}$ cars

$\underline{\qquad}$ x $\underline{\qquad}$ = $\underline{\qquad}$

4. $\frac{1}{6}$ of 12 fish

$\underline{\qquad}$ ÷ $\underline{\qquad}$ = $\underline{\qquad}$ fish

$\underline{\qquad}$ x $\underline{\qquad}$ = $\underline{\qquad}$

---

7 ⟩ Practice · · · · · · · **Problem Solving with Fractions**

Division can be used to find the fraction amount in word problems.

Step 1: Read the word problem and circle the important information.

Sandy had ⑥ pieces of candy. She ate $\frac{1}{2}$ of the candy. How many pieces of candy did Sandy eat?

Step 2: Write the math problem.

6 ÷ 2 =

Step 3: Solve the math problem.

6 ÷ 2 = 3

Sandy ate 3 pieces of candy.

Step 4: Use multiplication to check the answer.

3 x 2 = 6

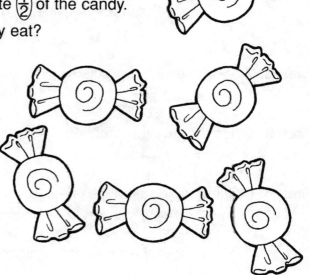

Directions: Read and solve each word problem. Circle the important information. Use this information to solve the problem. Remember to show your work. The first problem has already been done for you.

1. Joe ordered a jumbo, 16-slice pizza. Joe ate $\frac{1}{4}$ of the pizza. How many pieces of pizza did Joe eat? $16 \div 4 = 4$ $4 \times 4 = 16$ Joe ate ____4____ pieces of pizza.	**2.** Marisa has 21 yards of fabric. She used $\frac{1}{3}$ of the fabric to make some dresses. How many yards of fabric did Marisa use? Marisa used _____ yards of fabric.
3. Deanna made 24 cookies. Her dog, Spot, ate $\frac{1}{6}$ of the cookies. How many cookies did Spot eat? Spot ate _____ cookies.	**4.** Joshua planted 30 carrot seeds. Only $\frac{1}{5}$ of them sprouted. How many carrot seeds sprouted? Only _____ carrot seeds sprouted.
5. There were 18 gallons of gas in the car. We used $\frac{1}{9}$ of the gas to drive to the mall. How many gallons of gas did we use? We used _____ gallons of gas.	**6.** Bobby had 20 erasers. He gave his brother Benny $\frac{1}{10}$ of the erasers. How many erasers did Bobby give to Benny? Bobby gave Benny _____ erasers.

Multiplication and division can be used to find the fraction amount in word problems.

Step 1: Read the word problem and circle the important information.

Snow White had ⑭ apples. She ate $\frac{2}{7}$ of the apples.
How many apples did Snow White eat?

Step 2: Write the division problem.

14 ÷ 7 =

Step 3: Solve the division problem.

14 ÷ 7 = 2

Step 4: Multiply the quotient (answer) by the numerator (2).

2 x 2 = 4

Snow White ate 4 of the apples.

Directions: Read and solve each word problem. Circle the important information. Use the information to write the math problems. Remember to show your work. The first problem has already been done for you.

1. Rapunzel's hair is ⑱ feet long. The prince climbed up $\frac{2}{3}$ of the length of her hair. How many feet did the prince climb? 18 ÷ 3 = 6 6 x 2 = 12 The prince climbed __12__ feet.	**2.** The three bears collected 20 pails of honey. Papa Bear ate $\frac{9}{10}$ of the honey. How many pails of honey did Papa Bear eat? Papa Bear ate _____ pails of honey.
3. Cinderella tried on 25 pairs of slippers. Only $\frac{3}{5}$ of the slippers fit. How many pairs of slippers fit Cinderella's feet? _____ pairs of slippers fit.	**4.** Hansel and Gretel had 50 pieces of bread. They used $\frac{2}{5}$ of the pieces to mark their path. How many pieces of bread did they use? They used _____ pieces of bread.
5. Little Red Riding Hood carried 27 jars of jam to her grandmother's house. She broke $\frac{2}{3}$ of the jars running from the wolf. How many jars of jam broke? _____ jars of jam broke.	**6.** The troll charges to cross his bridge. Out of 28 animals who crossed his bridge, only $\frac{2}{7}$ paid the toll. How many animals paid the toll? _____ animals paid the toll.

Learning Notes

In this unit, children will convert fractions to decimals using a calculator when necessary. They will compare fractions and decimals and put them in order, smallest to greatest.

Materials

- calculators
- scratch paper
- pencils

Teaching the Lesson

Model for the children the relationship between fractions and decimals.

$\frac{2}{10}$ is the same as .2	$\frac{16}{100}$ is the same as .16
(two tenths)　　　　(two tenths)	(sixteen hundredths)　　(sixteen hundredths)

On a piece of scratch paper, have the children convert the following fractions to decimals:

$\frac{3}{100}$ (.03)　　　　$\frac{15}{100}$ (.15)　　　　$\frac{39}{100}$ (.39)

$\frac{7}{10}$ (.7)　　　　$\frac{4}{10}$ (.4)　　　　$\frac{27}{100}$ (.27)

On a piece of scratch paper, have the children convert the following decimals to fractions:

.58 ($\frac{58}{100}$)　　　　.44 ($\frac{44}{100}$)　　　　.78 ($\frac{78}{100}$)

.3 ($\frac{3}{10}$)　　　　.8 ($\frac{8}{10}$)　　　　.63 ($\frac{63}{100}$)

Model for the children how to convert other fractions to decimals with the use of a calculator.

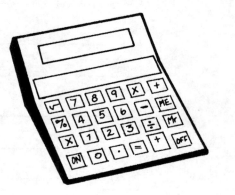

Using a Calculator

Step 1: Enter the numerator.

Step 2: Press $\div$.

Step 3: Enter the denominator.

Step 4: Press $=$.

Step 5: Record the answer.

Have the children convert the following fractions to decimals. Round off to the nearest hundredths place.

$\frac{1}{7}$ (.14)　　　　$\frac{2}{3}$ (.67)　　　　$\frac{7}{8}$ (.88)

Fractions can also be written as decimals.

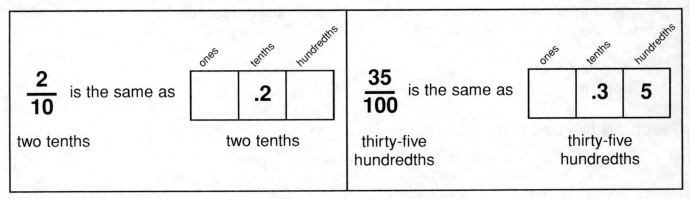

$\dfrac{2}{10}$ is the same as

ones	tenths	hundredths
	.2	

two tenths   two tenths

$\dfrac{35}{100}$ is the same as

ones	tenths	hundredths
	.3	5

thirty-five
hundredths

thirty-five
hundredths

Directions: Draw a line matching each fraction in the middle column to its decimal equivalent and word equivalent. The first one has already been done for you.

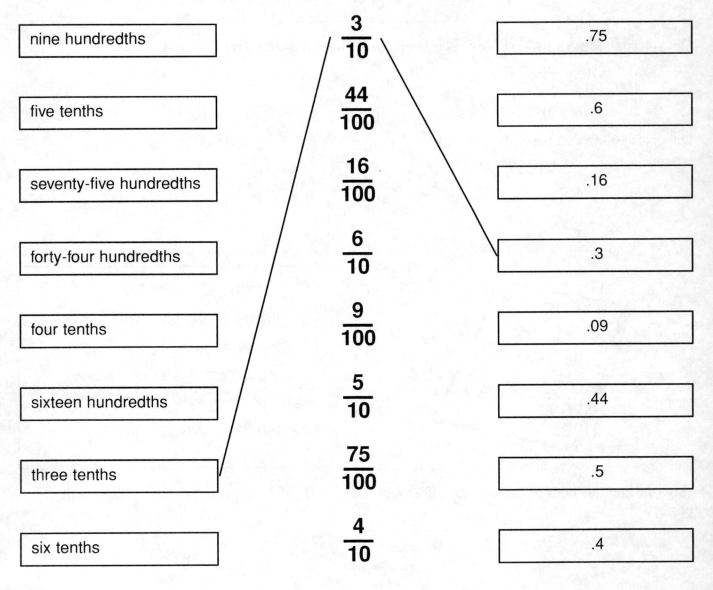

nine hundredths

five tenths

seventy-five hundredths

forty-four hundredths

four tenths

sixteen hundredths

three tenths

six tenths

$\dfrac{3}{10}$

$\dfrac{44}{100}$

$\dfrac{16}{100}$

$\dfrac{6}{10}$

$\dfrac{9}{100}$

$\dfrac{5}{10}$

$\dfrac{75}{100}$

$\dfrac{4}{10}$

.75

.6

.16

.3

.09

.44

.5

.4

Directions: Look at each set of fractions and decimals. Circle the larger one. The first one has already been done for you.

1. (7/10) .4	2. 22/100 .46	3. 79/100 .78
4. 8/10 .6	5. 98/100 .17	6. 9/10 .2
7. 5/10 .8	8. 56/100 .60	9. 3/10 .1
10. 35/100 .34	11. 41/100 .95	12. 2/10 .5
13. 4/10 .7	14. 63/100 .23	15. 1/10 .3
16. 84/100 .52	17. 15/100 .81	18. 79/100 .78

Directions: Write the following sets of fractions and decimals in order from smallest to largest.

19. (.4 $\frac{2}{10}$ $\frac{3}{10}$ .5) _____, _____, _____, _____

20. (.71 $\frac{65}{100}$ $\frac{83}{100}$ .59) _____, _____, _____, _____

21. (.7 $\frac{1}{10}$ $\frac{6}{10}$ .8) _____, _____, _____, _____

22. (.57 $\frac{46}{100}$ .99 $\frac{39}{100}$) _____, _____, _____, _____

A calculator can be used in converting (changing) fractions to decimals. Look at the example below.

Find the decimal for $\frac{3}{8}$ using a calculator.

Step 1: Enter the numerator $\boxed{3}$.

Step 2: Press $\boxed{\div}$.

Step 3: Enter the denominator $\boxed{8}$.

Step 4: Press $\boxed{=}$.

Step 5: Record the answer.
$\frac{3}{8}$ = .375

Step 6: Round off the number to the nearest hundredths place.
.375 = .38

Directions: Use a calculator to convert each fraction below to a decimal. Write the number to two decimal places in each box. Round off the numbers if necessary. Decode the secret message by writing the letter that goes with each decimal on the line at the bottom of the page.

A	C	E	F	I	N	O	R	S	T	U
$\frac{5}{6}$	$\frac{4}{5}$	$\frac{8}{9}$	$\frac{2}{6}$	$\frac{5}{7}$	$\frac{3}{5}$	$\frac{1}{4}$	$\frac{6}{7}$	$\frac{3}{8}$	$\frac{7}{8}$	$\frac{4}{9}$

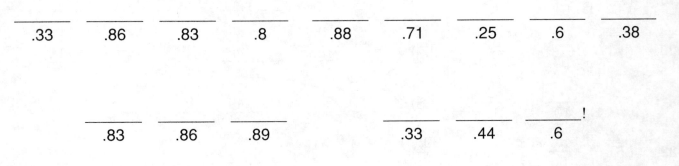

___ ___ ___ ___   ___ ___ ___   ___ ___
.33 .86 .83 .8   .88 .71 .25   .6   .38

___ ___ ___   ___ ___ ___!
.83 .86 .89   .33 .44 .6

Learning Notes

In this unit, children will convert money to fractions and fractions to money. They will compare money and fractions using the symbols > (greater than), < (less than), and = (equal to). They will also estimate the amount of money for a specific fraction.

Teaching the Lesson

Fraction "Cents" (page 38)

Model for the children the relationship between money (cents) and fractions. Remind the children that it takes 100 cents to make $1.00.

Money	⟶	Fraction	Money	⟶	Fraction
9¢ = $0.09 (9 hundredths)		$\dfrac{9 \text{ (cents)}}{100 \text{ (cents)}}$	57¢ = $0.57 (57 hundredths)		$\dfrac{57 \text{ (cents)}}{100 \text{ (cents)}}$

Money and Fractions (page 39)

Model for the children how dollars and cents can be written as a mixed number and how a mixed number can be converted to money.

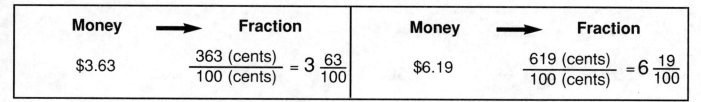

Money	⟶	Fraction	Money	⟶	Fraction
$3.63		$\dfrac{363 \text{ (cents)}}{100 \text{ (cents)}} = 3\dfrac{63}{100}$	$6.19		$\dfrac{619 \text{ (cents)}}{100 \text{ (cents)}} = 6\dfrac{19}{100}$

Estimating (page 40)

Discuss with the children how estimating is a reasonable guess. For example, $\frac{1}{2}$ of $15 is about $7 or $8. ($7 + $7 = $14, $8 + $8 = $16). To estimate $\frac{1}{4}$ of $15, do the following.

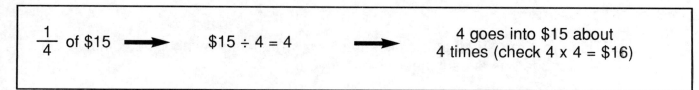

$\dfrac{1}{4}$ of $15 ⟶ $15 ÷ 4 = 4 ⟶ 4 goes into $15 about 4 times (check 4 x 4 = $16)

Money can be written as a fraction.

1¢	is the same as	$\dfrac{1}{100}$	$0.41	is the same as	$\dfrac{41}{100}$

It takes 100 cents to make a dollar. One penny is 1 one-hundredth of a dollar.

It takes 100 cents to make a dollar. 41¢ is 41 hundredths of a dollar.

Directions: Write each amount of money as a fraction. The first one has already been done for you.

1.

$9¢ = \dfrac{9}{100}$

2.

$44¢ = \underline{}$

3.

$98¢ = \underline{}$

4.

$65¢ = \underline{}$

5.

$39¢ = \underline{}$

6.

$27¢ = \underline{}$

7.

$0.73 = \underline{}$

8.

$0.88 = \underline{}$

9.

$0.23 = \underline{}$

10.

$0.15 = \underline{}$

11.

$0.50 = \underline{}$

12.

$0.11 = \underline{}$

Directions: Use the > (greater than), < (less than), or = (equal to) symbols to compare the numbers. The first one has already been done for you.

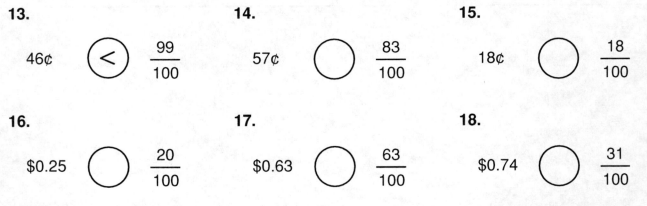

13.

$46¢ \enspace \boxed{<} \enspace \dfrac{99}{100}$

14.

$57¢ \enspace \bigcirc \enspace \dfrac{83}{100}$

15.

$18¢ \enspace \bigcirc \enspace \dfrac{18}{100}$

16.

$0.25 \enspace \bigcirc \enspace \dfrac{20}{100}$

17.

$0.63 \enspace \bigcirc \enspace \dfrac{63}{100}$

18.

$0.74 \enspace \bigcirc \enspace \dfrac{31}{100}$

Dollars and cents can be written as fractions too.

$1.00 is the same as $\dfrac{100}{100}$ or $\dfrac{1}{1}$ or 1	**$1.23** is the same as $\dfrac{123}{100}$ or $1\dfrac{23}{100}$

Directions: Write each amount of money as a mixed number. The first one has already been done for you.

1.

$5.81 $= 5\dfrac{81}{100}$

2.

$2.71 $=$ ——

3.

$3.19 $=$ ——

4.

$1.86 $=$ ——

5.

$4.04 $=$ ——

6.

$5.11 $=$ ——

7.

$1.63 $=$ ——

8.

$2.10 $=$ ——

9.

$3.07 $=$ ——

10.

$4.29 $=$ ——

11.

$7.99 $=$ ——

12.

$3.00 $=$ ——

Directions: Convert each fraction to money. The first one has already been done for you.

13.

$6\dfrac{19}{100}$ = 6.19

14.

$2\dfrac{1}{100}$ =

15.

$4\dfrac{98}{100}$ =

16.

$3\dfrac{68}{100}$ =

17.

$1\dfrac{50}{100}$ =

18.

$2\dfrac{27}{100}$ =

Estimation can be used when working with money and fractions. Look at the example below.

Step 1: Read the word problem.	Step 2: Circle the estimated dollar amount.
Janet had $13 to spend at the toy store. She spent $\frac{1}{2}$ of her money on a board game. About how much money does Janet have left?	(*Think:* One half of thirteen is about $6. $6 + $6 = $12. Janet has about $6 left after buying the board game.) $4 ($6) $10

Directions: Read each word problem. Circle the estimated dollar amount.

1. Ken spent $11 on school supplies. If $\frac{1}{3}$ of the money was used to buy crayons, about how much money did Ken spend on crayons?

    $2 $4 $6

2. Hannah earned $14 by cleaning the house and the yard. If $\frac{2}{3}$ of the money was earned by raking leaves, about how much money did Hannah earn by raking the leaves?

    $5 $7 $9

3. Oliver spent $10 at the movies. If $\frac{2}{3}$ of the money was used to buy his ticket, about how much money did the ticket cost?

    $5 $7 $9

4. Belle paid $\frac{1}{4}$ of the $22 price tag to put the tent on hold, about how much was Belle's deposit?

    $3 $5 $7

5. Cody cleaned 5 windows for $16. Cody was paid the same amount for cleaning each window ($\frac{1}{5}$ of $16). About how much did Cody earn cleaning each window?

    $3 $4 $5

6. Ciera had $9 in her piggy bank. She spent $\frac{3}{4}$ of the money on a book. About how much money did the book cost?

    $6 $9 $12

Directions: Solve each problem. Then circle the answer that is worth more. The first one has been done for you.

7. $\frac{1}{3}$ of $12 = ____$4____ $\frac{1}{4}$ of $20 = ____($5)____

8. $\frac{1}{2}$ of $8 = _____ $\frac{1}{3}$ of $9 = _____

9. $\frac{1}{4}$ of $16 = _____ $\frac{1}{2}$ of $10 = _____

10. $\frac{1}{5}$ of $30 = _____ $\frac{1}{4}$ of $28 = _____

When comparing fractions with different denominators, look at the total shaded area for each shape.

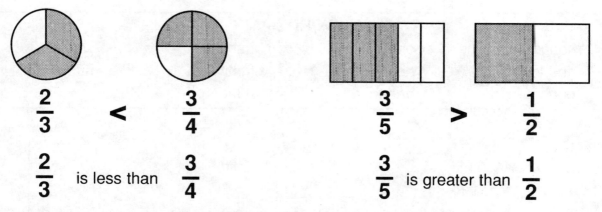

$$\frac{2}{3} < \frac{3}{4}$$

$$\frac{2}{3} \text{ is less than } \frac{3}{4}$$

$$\frac{3}{5} > \frac{1}{2}$$

$$\frac{3}{5} \text{ is greater than } \frac{1}{2}$$

Directions: Color in each shape to show the correct fraction. Compare the two fractions and then complete each fraction sentence. The first one has already been done for you.

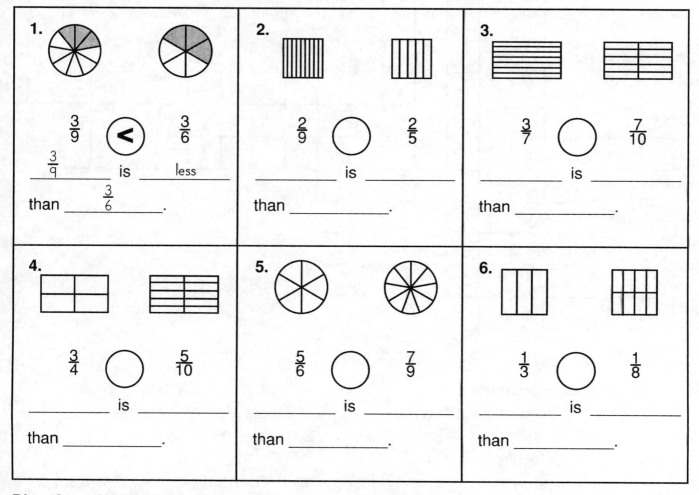

1. $\frac{3}{9}$ < $\frac{3}{6}$

$\frac{3}{9}$ is _less_ than $\frac{3}{6}$.

2. $\frac{2}{9}$ ◯ $\frac{2}{5}$

_____ is _____ than _____ .

3. $\frac{3}{7}$ ◯ $\frac{7}{10}$

_____ is _____ than _____ .

4. $\frac{3}{4}$ ◯ $\frac{5}{10}$

_____ is _____ than _____ .

5. $\frac{5}{6}$ ◯ $\frac{7}{9}$

_____ is _____ than _____ .

6. $\frac{1}{3}$ ◯ $\frac{1}{8}$

_____ is _____ than _____ .

Directions: Write the fractions in order from smallest to largest.

7. $\left(\frac{3}{4}, \frac{1}{4}, \frac{2}{4}\right)$ _____, _____, _____

8. $\left(\frac{4}{6}, \frac{1}{6}, \frac{3}{6}\right)$ _____, _____, _____

Directions: Use the words in the Word Bank to help you complete the crossword puzzle. Read each clue and write the answer in the correct boxes.

WORD BANK

congruent	one half	fraction	whole
one fourth	equivalent	tenth	numerator
denominator	one third	hundredth	

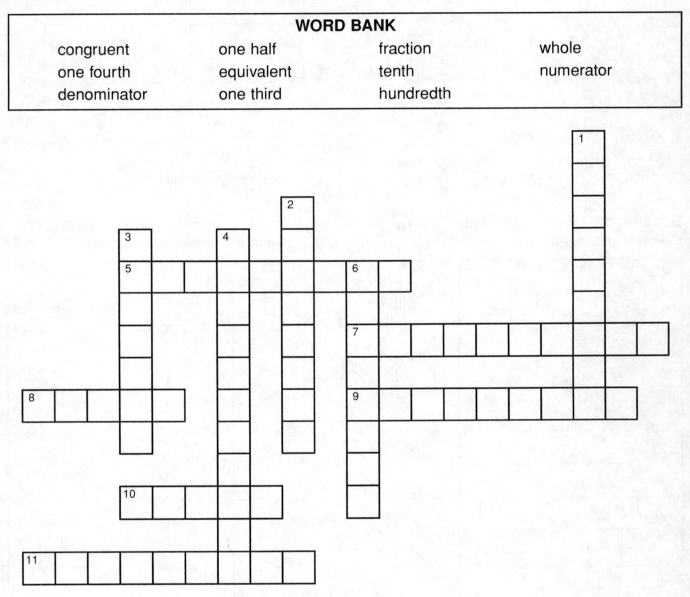

Across

5. the top number of a fraction

7. same fraction as another

8. a complete circle is a _____ circle

9. .01

10. .1

11. a quarter

Down

1. same size and shape

2. part of a whole

3. 1/2

4. the bottom number of a fraction

6. 1/3

Directions: Use a calculator to convert each fraction to a decimal. Write each decimal in the space next to the fraction. Then answer the questions on page 44.

	1	2	3	4	5	6	7	8	9	10
10	$\frac{1}{10}=$	$\frac{2}{10}=$	$\frac{3}{10}=$	$\frac{4}{10}=$	$\frac{5}{10}=$	$\frac{6}{10}=$	$\frac{7}{10}=$	$\frac{8}{10}=$	$\frac{9}{10}=$	$\frac{10}{10}=$
9	$\frac{1}{9}=$	$\frac{2}{9}=$	$\frac{3}{9}=$	$\frac{4}{9}=$	$\frac{5}{9}=$	$\frac{6}{9}=$	$\frac{7}{9}=$	$\frac{8}{9}=$	$\frac{9}{9}=$	
8	$\frac{1}{8}=$	$\frac{2}{8}=$	$\frac{3}{8}=$	$\frac{4}{8}=$	$\frac{5}{8}=$	$\frac{6}{8}=$	$\frac{7}{8}=$	$\frac{8}{8}=$		
7	$\frac{1}{7}=$	$\frac{2}{7}=$	$\frac{3}{7}=$	$\frac{4}{7}=$	$\frac{5}{7}=$	$\frac{6}{7}=$	$\frac{7}{7}=$			
6	$\frac{1}{6}=$	$\frac{2}{6}=$	$\frac{3}{6}=$	$\frac{4}{6}=$	$\frac{5}{6}=$	$\frac{6}{6}=$				
5	$\frac{1}{5}=$	$\frac{2}{5}=$	$\frac{3}{5}=$	$\frac{4}{5}=$	$\frac{5}{5}=$					
4	$\frac{1}{4}=$	$\frac{2}{4}=$	$\frac{3}{4}=$	$\frac{4}{4}=$						
3	$\frac{1}{3}=$	$\frac{2}{3}=$	$\frac{3}{3}=$							
2	$\frac{1}{2}=$	$\frac{2}{2}=$								
1	$\frac{1}{1}=$									

Directions: Using crayons, color the fractions with specific decimals on page 43 in the following colors:

Red: decimals of 1	**Green:** decimals of .33	**Brown:** decimals of .4
Orange: decimals of .5	**Blue:** decimals of .67	**Gray:** decimals of .6
Yellow: decimals of .25	**Purple:** decimals of .75	**Black:** decimals of .8

Directions: Answer the following questions after completing the Fractions to Decimals Table on page 43.

1. What do you notice about the fractions for each color group of decimals?

_____

_____

2. What kind of pattern is made when the decimals of 1 were colored?

_____

_____

3. Which fractions had decimals less than .5?

_____

_____

4. Which fractions had decimals greater than .5?

_____

_____

5. When looking at the columns on the Fractions to Decimals Table, what do you notice about the decimals going from the top of each column to the bottom of each column?

_____

_____

6. Do you see any other patterns in the Fractions to Decimals Table? If yes, tell about the pattern you found.

_____

_____

Materials

- computer with paint program
- printer
- file with drawings shown below
- chart models of fractions (optional)

Before the Computer

- The teacher should create and save a file that contains the fractions and shapes shown below.

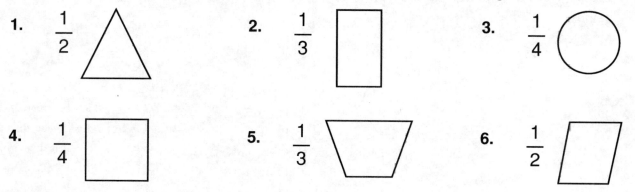

- The teacher should review fractions with the children by reminding them that they are parts of a whole. Use chart models of fractions to review the common fractions of $\frac{1}{4}$, $\frac{1}{3}$, and $\frac{1}{2}$. The teacher can also use manipulatives for the children to make fractions of a whole.
- The children should be familiar with the draw and paint tools in the paint software program being used.
- The children should know how to print their work.

At the Computer

- Display the paint file with pictures of the shapes above on the monitor.
- Tell the children they will divide these shapes into different fractions.
- Ask the children to divide the first picture in half using the line tool to draw the line.
- Show the children how to use the floodfill tool (paint bucket icon in most paint programs) to fill in the fractions with different colors.
- Ask the children to divide the next picture into thirds with the line tool.
- Have the children finish the pictures in a file of their own.
- Ask the children to print the file after completing the activity.
- Remind the children to close the file without saving the changes. This restores the teacher's file to its original state.

Extensions

- More painting files can be added to the folder to allow students new opportunities for making fractions.
- Sets of objects can be placed in the file to divide into fractions.
- Children can create their own shapes for other children to make into fractions.

Answer Key

Page 6

1. 1/4
2. 2/3
3. 1/2
4. 3/8
5. 7/10
6. 1/3
7. 3/4
8. 2/4

Page 7

1. 2/6
2. 1/3
3. 2/4
4. 2/3
5. 3/5
6. 4/8
7. 6/8
8. 1/4

Page 8

1. 1/3
2. 4/6
3. 2/5
4. 3/4
5. 1/2
6. 1/2
7. 5/9
8. 1/4
9. 3/4
10. 3/6
11. 2/4
12. 2/3

Page 10

1. 3
2. 4
3. 2
4. 8
5. 5
6. 4
7. 8
8. 8
9. 3

Page 11

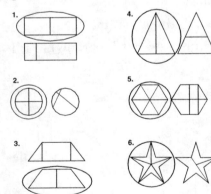

7. Answers will vary.
8. The parts of the shape are the same size and the same shape.

Page 12

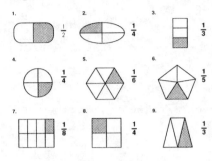

*The above is just one way.

Page 14

1. 8 bats
2. 8, 4
3. 4/8
4. 2 balls
5. 2, 1
6. 1/2
7. 4 cats
8. 4, 2
9. 2/4
10. 6 keys
11. 6, 3
12. 3/6
13. 2 birds
14. 2, 1
15. 1/2
16. 8 stars
17. 8, 4
18. 4/8

Page 15

1. 9 bones; 9, 3; 3/9
2. 6 books; 6, 2; 2/6
3. 3 ice cream cones; 3, 1; 1/3
4. 12 apples; 12, 4; 4/12

Page 16

1. 1/5 of 10 is 2.
   Check answer:
   2 + 2 + 2 + 2 + 2 = 10
2. 1/2 of 10 is 5.
   Check answer:
   5 + 5 = 10
3. 1/4 of 8 is 2.
   Check answer:
   2 + 2 + 2 + 2 = 8
4. 1/8 of 8 is 1.
   Check answer:
   1 + 1 + 1 + 1 + 1 + 1 + 1 + 1 = 8

Page 18

1. 1, 2, 3, 4, 5, 6
2. one whole, one half, one third, one fourth, one fifth, one sixth
3. 1/1, 1/2, 1/3, 1/4, 1/5, 1/6

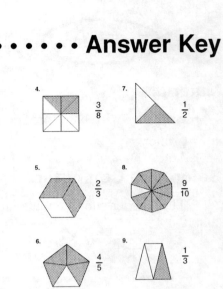

Page 19

1. 1/3 < 2/3; 1/3 is less than 2/3.
2. 3/4 > 2/4; 3/4 is greater than 2/4.
3. 3/6 < 5/6; 3/6 is less than 5/6.
4. 1/4 < 3/4; 1/4 is less than 3/4.
5. 4/7 < 5/7; 4/7 is less than 5/7.
6. 4/5 > 1/5; 4/5 is greater than 1/5.

Page 20

1/10, 1/9, 1/8, 1/7, 1/6, 1/5, 1/4, 1/3, 1/2

Page 22

1. whole
2. 4 pears
3. whole
4. 3 strawberries
5. mixed
6. 3 1/2 oranges
7. whole
8. 1 watermelon
9. mixed
10. 1 1/2 pineapples
11. mixed
12. 2 1/2 peaches
13. 1, 1 1/2, 2 1/2, 3, 3 1/2, 4

Page 23

1. 1 1/2
2. 3 1/2
3. 1 1/3
4. 1 1/4

Page 24

1. 4 2/3
2. 3 1/2
3. 2 1/3
4. 1 3/4
5. 1 3/4, 2 1/3, 3 1/2, 4 2/3

Page 26

1. $\frac{1}{2}$ $\frac{2}{4}$ 2. $\frac{1}{2}$ $\frac{3}{6}$

3. $\frac{1}{2}$ $\frac{2}{4}$ 4. $\frac{1}{2}$ $\frac{4}{8}$

5. The two shapes have the same total amount shaded.

Page 27

1. 16/18
2. 24/56
3. 21/24
4. 32/40
5. 7/14
6. 15/30
7. 36/42
8. 36/45
9. 8/40
10. The fraction stays the same.

Page 28

1. 3/4 x 8/8 = 24/32
   5/8 x 4/4 = 20/32
   no

2. 2/3 x 9/9 = 18/27
   6/9 x 3/3 = 18/27
   yes

3. 1/3 x 6/6 = 6/18
   2/6 x 3/3 = 6/18
   yes

4. 1/2 x 3/3 = 3/6
   2/3 x 2/2 = 4/6
   no

5. 2/3 x 5/5 = 10/15
   3/5 x 3/3 = 9/15
   no

6. 1/4 x 5/5 = 5/20
   2/5 x 4/4 = 8/20
   no

Page 30

1. 16 ÷ 4 = 4 buttons; 4 x 4 = 16
2. 15 ÷ 3 = 5 stars; 5 x 3 = 15
3. 10 ÷ 5 = 2 cars; 2 x 5 = 10
4. 12 ÷ 6 = 2 fish; 2 x 6 = 12

Page 31

1. 16 ÷ 4 = 4; 4 x 4 = 16; Joe ate 4 pieces of pizza.
2. 21 ÷ 3 = 7; 7 x 3 = 21; Marisa used 7 yards of fabric.
3. 24 ÷ 6 = 4; 4 x 6 = 24; Spot ate 4 cookies.
4. 30 ÷ 5 = 6; 6 x 5 = 30; Only 6 carrot seeds sprouted.
5. 18 ÷ 9 = 2; 2 x 9 = 18; We used 2 gallons of gas.
6. 20 ÷ 10 = 2; 2 x 10 = 20; Bobby gave Benny 2 erasers.

Page 32

1. 18 ÷ 3 = 6; 6 x 2 = 12; The prince climbed 12 feet.
2. 20 ÷ 10 = 2; 2 x 9 = 18; Papa Bear ate 18 pails of honey.
3. 25 ÷ 5 = 5; 5 x 3 = 15; 15 pairs of slippers fit.
4. 50 ÷ 5 = 10; 10 x 2 = 20; They used 20 pieces of bread.
5. 27 ÷ 3 = 9; 9 x 2 = 18; 18 jars of jam broke.
6. 28 ÷ 7 = 4; 4 x 2 = 8; 8 animals paid the toll.

Page 34

3/10, three tenths, .3

44/100, forty-four hundredths, .44

16/100, sixteen hundredths, .16

6/10, six tenths, .6

9/100, nine hundredths, .09

5/10, five tenths, .5

75/100, seventy-five hundredths, .75

4/10, four tenths, .4

Page 35

1. 7/10
2. .46
3. 79/100
4. 8/10
5. 98/100
6. 9/10
7. .8
8. .60
9. 3/10
10. 35/100
11. .95
12. .5
13. .7
14. 63/100
15. .3
16. 84/100
17. .81
18. 79/100
19. 2/10, 3/10, .4, .5
20. .59, 65/100, .71, 83/100
21. 1/10, 6/10, .7, .8
22. 39/100, 46/100, .57, .99

Page 36

Chart: .83, .8, .89, .33, .71, .6, .25, .86, .38, .88, .44

Message: Fractions are fun!

Page 38

1. 9/100
2. 44/100
3. 98/100
4. 65/100
5. 39/100
6. 27/100
7. 73/100
8. 88/100
9. 23/100
10. 15/100
11. 50/100
12. 11/100
13. 46¢ < 99/100
14. 57¢ < 83/100
15. 18¢ = 18/100
16. $0.25 > 20/100
17. $0.63 = 63/100
18. $0.74 > 31/100

Page 39

1. 5 81/100
2. 2 71/100
3. 3 19/100
4. 1 86/100
5. 4 4/100
6. 5 11/100
7. 1 63/100
8. 2 10/100
9. 3 7/100
10. 4 29/100
11. 7 99/100
12. 3 0/100 or 3
13. $6.19
14. $2.01
15. $4.98
16. $3.68
17. $1.50
18. $2.27

Page 40

1. $4
2. $9
3. $7
4. $5
5. $3
6. $6
7. $4, ($5)
8. ($4) $3
9. $4, ($5)
10. $6, ($7)

Page 41

1. 3/9 < 3/6

   3/9 is less than 3/6

2. 2/9 < 2/5

   2/9 is less than 2/5

3. 3/7 < 7/10

   3/7 is less than 7/10

4. 3/4 > 5/10

   3/4 is greater than 5/10

5. 5/6 > 7/9

   5/6 is greater than 7/9

6. 1/3 > 1/8

   1/3 is greater than 1/8

7. 1/4, 2/4, 3/4

8. 1/6, 3/6, 4/6

Page 42

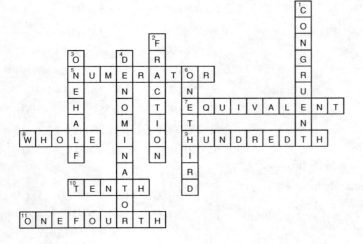

Page 43

.1, .2, .3, .4, .5, .6, .7, .8, .9, 1

.11, .22, .33, .44, .56, .67, .78, .89, 1

.13, .25, .38, .5, .63, .75, .88, 1

.14, .29, .43, .57, .71, .86, 1

.17, .33, .5, .67, .83, 1

.2, .4, .6, .8, 1

.25, .5, .75, 1

.33, .67, 1

.50, 1

1

Page 44

1. Sample answer: They are all equivalent fractions.

2. Sample answer: The decimals of 1 make a stair pattern.

3. 1/10, 2/10, 3/10, 4/10, 1/9, 2/9, 3/9, 4/9, 1/8, 2/8, 3/8, 1/7, 2/7, 3/7, 1/6, 2/6, 1/5, 2/5, 1/4, 1/3

4. 6/10, 7/10, 8/10, 9/10, 10/10, 5/9, 6/9, 7/9, 8/9, 9/9, 5/8, 6/8, 7/8, 8/8, 4/7, 5/7, 6/7, 7/7, 4/6, 5/6, 6/6, 3/5, 4/5, 5/5, 3/4, 4/4, 2/3, 3/3, 2/2, 1/1 or 1

5. Sample answer: The decimals become larger.

6. Answers will vary.

Page 45

Answers will vary.